NATURAL DISASTERS
MEETING THE CHALLENGE

VOLCANIC ERUPTION READINESS

Heather C. Hudak

Crabtree Publishing Company
www.crabtreebooks.com

CRABTREE
PUBLISHING COMPANY
WWW.CRABTREEBOOKS.COM

Author: Heather C. Hudak

Series research and development: Janine Deschenes, Reagan Miller

Editorial director: Kathy Middleton

Editor: Ellen Rodger

Proofreader: Roseann S. Biederman

Design and photo research: Katherine Berti

Images:

Alamy: Sijori Images/ ZUMAPRESS.com/ Alamy Live News: p. 38
Flickr: Charles G. Haacker: p. 27 (bottom left) Mr G's Travels: p. 21
iStockphoto: p. 13 (bottom), 18
NASA: p. 6 (center left)
Shutterstock:
Dewi Putra: p. 15
J. Helgason: p. 23
rui vale sousa: p. 37
The Mariner 2392: front cover (top), p. 22 (top)
U.S. Geological Survey:
p. 32 (bottom)
C. Oppenheimer: p. 26 (bottom)
Dave Harlow: p. 7
Hawaiian Volcano Observatory (HVO): p. 10 (top)
Jeffrey Marso: p. 32 (center)
Joe Bard: p. 29 (bottom)
Liz Westby: p. 26 (top)
Manny Nathensen: p. 31 (inset, top)
volcanoes.usgs.gov: Screen Shot 2020-02-04 at 4.43.51 PM: p. 31 (inset, bottom)
Wikimedia Commons:
Christian Bickel fingalo: p. 12
Geospatial Information Authority of Japan—UAV: p. 11 (top)
Hiroshige II: p. 6 (bottom right)
Mike Doukas—USGS Cascades Volcano Observatory: p. 27 (right)
The Getty Research Institute, 93-B15110: p. 19
Tim Evanson—Pierre-Jacques Volaire: p. 6 (center right)
U.S. Forest Service—Pacific Northwest Region: p. 10 (bottom)
USGS: p. 25 (top)

All other images by Shutterstock

Library and Archives Canada Cataloguing in Publication

Title: Volcanic eruption readiness / Heather C. Hudak.
Names: Hudak, Heather C., 1975- author.
Description: Series statement: Natural disasters: meeting the challenge | Includes bibliographical references and index.
Identifiers: Canadiana (print) 20190231459 | Canadiana (ebook) 20190231475 | ISBN 9780778774181 (softcover) | ISBN 9780778774075 (hardcover) | ISBN 9781427125088 (HTML)
Subjects: LCSH: Volcanic eruptions—Juvenile literature. | LCSH: Volcanoes—Juvenile literature. | LCSH: Emergency management—Juvenile literature. | LCSH: Hazard mitigation—Juvenile literature.
Classification: LCC QE522 .H83 2020 | DDC j363.34/95—dc23

Library of Congress Cataloging-in-Publication Data

Names: Hudak, Heather C., 1975- author.
Title: Volcanic eruption readiness / Heather C. Hudak.
Other titles: Natural disasters: meeting the challenge.
Description: New York, New York : Crabtree Publishing Company, [2020] | Series: Natural disasters: meeting the challenge | Includes bibliographical references and index.
Identifiers: LCCN 2019052903 (print) | LCCN 2019052904 (ebook) | ISBN 9780778774075 (hardcover) | ISBN 9780778774181 (paperback) | ISBN 9781427125088 (ebook)
Subjects: LCSH: Volcanoes--Juvenile literature. | Volcanic activity prediction--Juvenile literature. | Emergency management--Juvenile literature.
Classification: LCC QE527.5 H83 2020 (print) | LCC QE527.5 (ebook) | DDC 551.21--dc23
LC record available at https://lccn.loc.gov/2019052903
LC ebook record available at https://lccn.loc.gov/2019052904

Crabtree Publishing Company

www.crabtreebooks.com 1-800-387-7650

Printed in the U.S.A./042020/CG20200224

 In Canada: We acknowledge the financial support of the Government of Canada through the Canada Book Fund for our publishing activities.

Published in Canada
Crabtree Publishing
616 Welland Ave.
St. Catharines, Ontario
L2M 5V6

Published in the United States
Crabtree Publishing
PMB 59051
350 Fifth Avenue, 59th Floor
New York, New York 10118

Published in the United Kingdom
Crabtree Publishing
Maritime House
Basin Road North, Hove
BN41 1WR

Published in Australia
Crabtree Publishing
Unit 3–5 Currumbin Court
Capalaba
QLD 4157

Contents

INTRODUCTION

What a Blast!

Volcanoes are openings, or **vents**, in Earth's surface. They can erupt, spewing hot ash, **lava**, rocks, and gases high into the air and surrounding area. Named after Vulcan, the Roman god of fire, volcanoes have been shaping the planet since its early beginning billions of years ago. In fact, more than 80 percent of Earth's surface has been formed by volcanic activity. As volcanoes explode, they create mountains and craters. Lava that flows across the land breaks down over time to become fertile soils for farming.

Where Are They?

Volcanoes are found on every continent on Earth. No one knows for certain how many there are on the planet. Some scientists count each volcano vent as a single volcano. However, in a field of volcanoes there can be hundreds of vents fed by the same **magma** storage chamber. Some scientists count these as a single volcano.

Volcanoes can be found on other planets, too. Mars is home to the largest volcano in the solar system. Known as **OLYMPUS MONS,** *it spans 374 miles (624 km) wide and rises 16 miles (25 km) into the sky.*

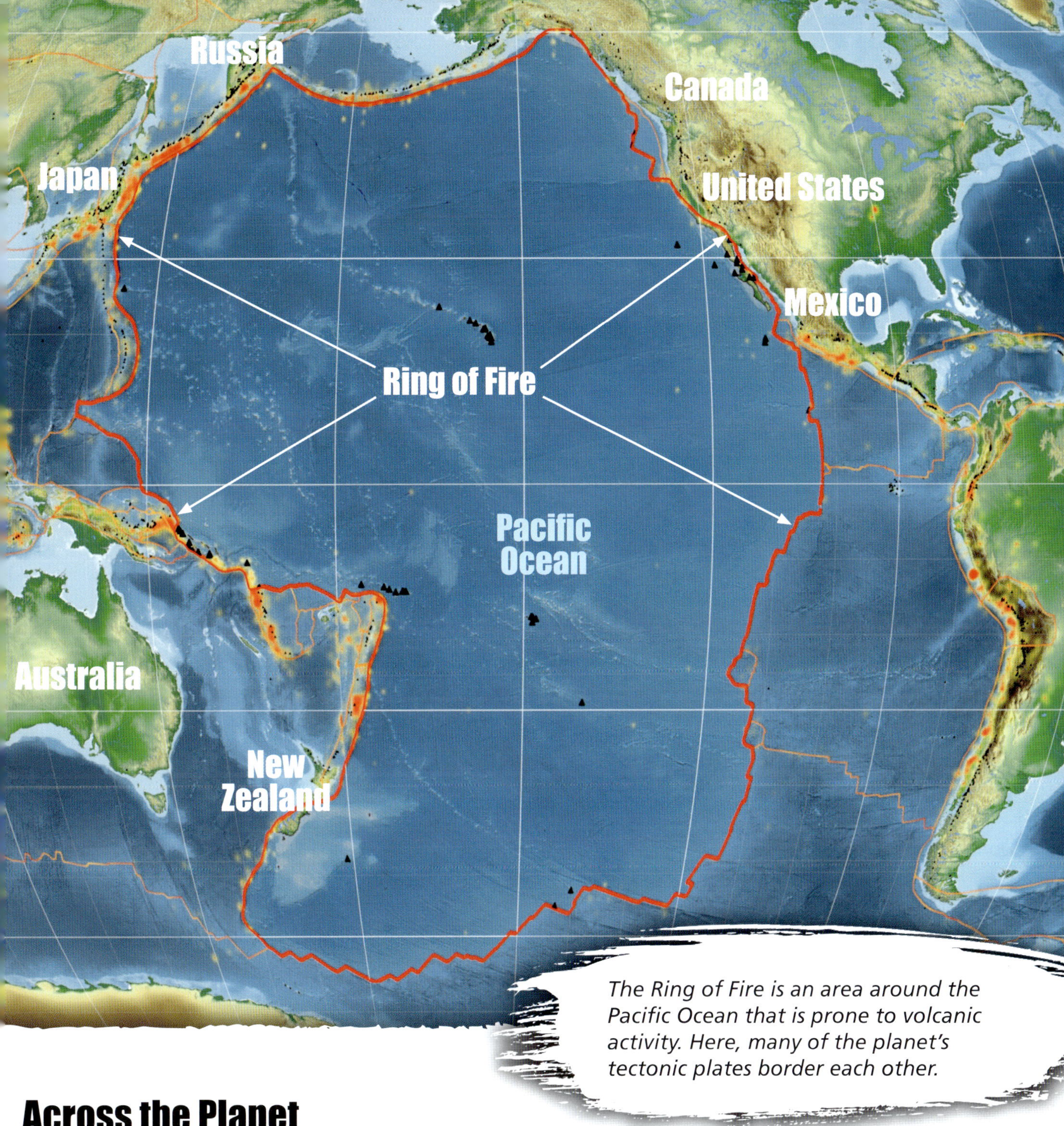

The Ring of Fire is an area around the Pacific Ocean that is prone to volcanic activity. Here, many of the planet's tectonic plates border each other.

Across the Planet

Approximately 1,500 volcanoes have erupted in the past 10,000 years and have the potential to erupt again. About 10 percent—or 161—of them are located within the United States, mainly in Hawaii, Oregon, California, Alaska, and Washington. There are at least 80 volcanoes under the ocean, too. The Ring of Fire is home to about 450, or 75 percent, of all active volcanoes in the world. This area forms a 24,900-mile (40,000 km) path along the boundaries of several **tectonic** plates near the Pacific Ocean.

When Volcanoes Erupt

There are about 50 to 60 volcanic eruptions on Earth each year. At any given time, there are about 20 volcanoes erupting. While eruptions are fairly common, many are not disruptive to humans. However, they can set off earthquakes, **tsunamis**, and rockfalls that claim lives and do a lot of damage.

Volcanic eruptions are natural events that have happened for billions of years. Sometimes, natural events are necessary for Earth to sustain life. For instance, when Earth formed, it had very little atmosphere. Erupting volcanoes spewed gases, such as water vapor and carbon dioxide. These created the atmosphere. As Earth cooled, the water vapor condensed to form Earth's oceans.

Natural events become natural disasters when they have an impact on humans and other life on Earth. By understanding natural events of the past, we can better predict and prevent natural disasters in the future. This helps reduce damage and loss of life.

Cleveland Volcano, Alaska

Mount Vesuvius, Italy

Kīlauea, Hawaii

Mount Asama, Japan

CASE STUDY
Mount Pinatubo

On June 15, 1991, Mount Pinatubo erupted in the Philippines. It was the largest eruption of the past century. Located on the island of Luzon, Mount Pinatubo is just 55 miles (90 km) from the nation's capital city, Manila. Three other major eruptions took place at the volcano about 500, 3,000, and 5,500 years ago. It was likely reawakened during a major earthquake in the region of Pinatubo in July 1990.

A small eruption on April 2 sent warning signs to the locals, and about 5,000 people were told to **evacuate** the area soon after. On June 5, an alert was issued that a major eruption was likely to come. Within a week nearly 60,000 people living in a 19-mile (30 km) radius of the volcano were told to leave their homes.

Mount Pinatubo began to erupt at 1:42 p.m. on June 15. The eruption lasted for nine hours. The **summit** of the volcano **collapsed** and caused multiple earthquakes in the area. At the time, a tropical storm was passing through the region. The heavy rains mixed with the ash to cover all of Luzon. Rooftops caved in from the weight of the ash.

The eruption likely impacted other natural events, such as flooding along the Mississippi River in 1993 and **drought** in Africa's Sahel region. The loss could have been much worse had it not been for the efforts of scientists from the U.S. Geological Survey (USGS) and the Philippine Institute of Volcanology and **Seismology** (PHIVOLCS). They began monitoring the volcano months in advance and worked with Filipino public officials to put evacuation plans in place well before the major eruption took place.

In total, the eruption of Mount Pinatubo and its aftermath killed about **800 PEOPLE** *and caused about* **$500 MILLION** *in damage.*

The eruption sent millions of tons (tonnes) of sulfur dioxide into the atmosphere. It increased the size of the ***ozone*** *hole over Antarctica and caused a decrease in the global climate that lasted about two years.*

CHAPTER 1

The Science Behind Volcanic Eruptions

For centuries, volcanoes have awed humans with their destructive force. In the past 500 years, about 100 volcanoes have erupted. Approximately 250,000 people have died due to volcanic eruptions since the 1700s. About two-thirds died in one of four major volcanic events: Mount Tambora in 1815, Krakatoa in 1883, Mount Pelee in 1902, and Mount Ruiz in 1975. Understanding the science behind volcanic eruptions can help save lives in the future.

Tectonic Activity

Earth's outermost layer, or crust, is made up of hard slabs of rock called tectonic plates. There are nine major plates and several smaller plates that form a sort of shell over Earth's **mantle**. The plates fit together like pieces of a jigsaw puzzle. They shift and glide over the mantle. Sometimes they even collide, break apart, or grind against one another. This interaction between the plates causes changes in Earth's surface. Mountains, ocean trenches, earthquakes, and volcanoes take place along the boundaries between plates.

VOLCANIC ERUPTION

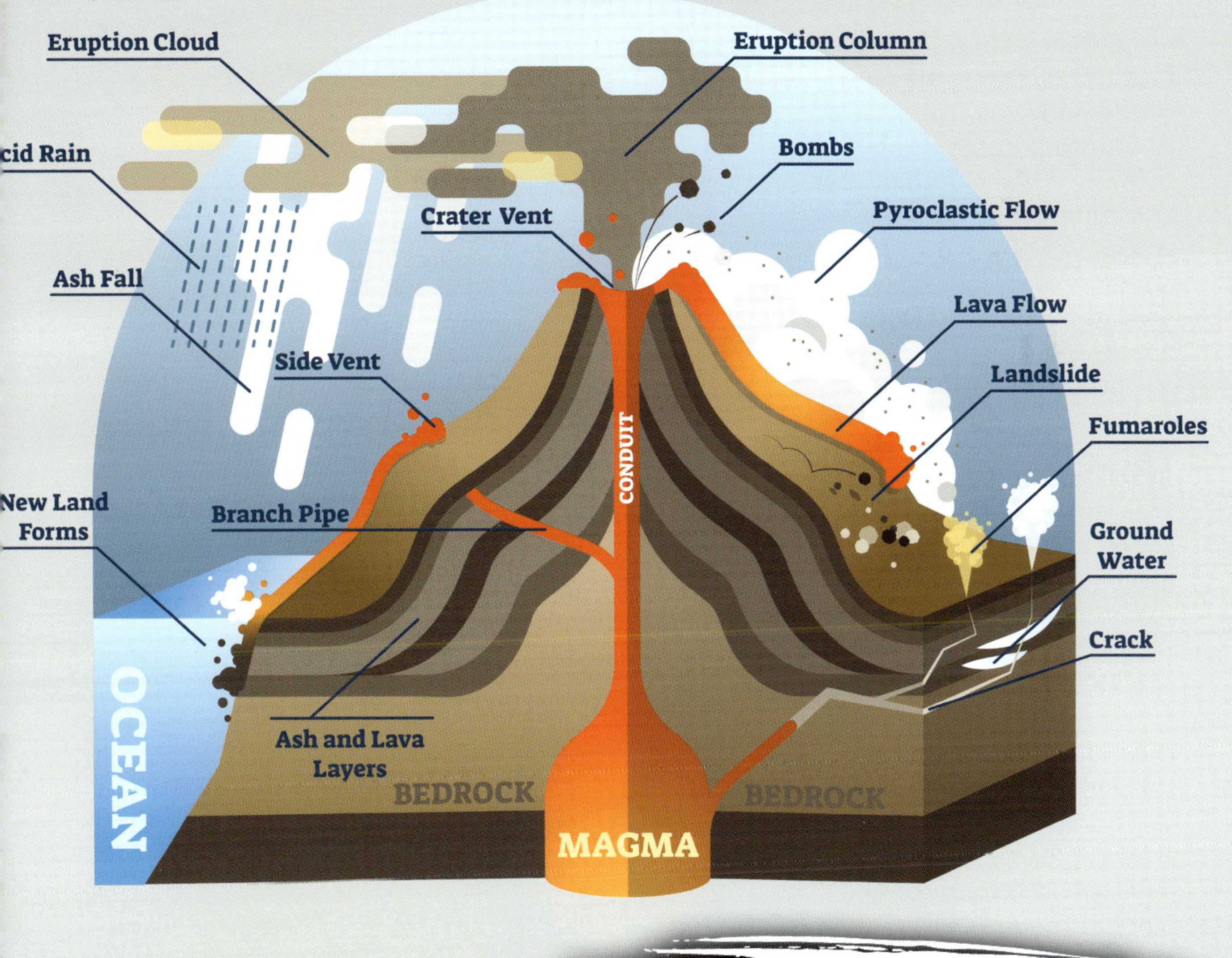

For many volcanoes, magma rises up a central conduit, or throat. So many things can happen during an eruption, from **pyroclastic flows**, *to vented gas fumaroles.*

How Volcanoes Erupt

Magma consists of molten rock, crystals, and dissolved gas. It is found many miles (km) beneath Earth's surface where it is so hot that rocks begin to melt. Magma is lighter than the surrounding rock, which is still in a solid form. It floats on top of the solid rock as it rises toward Earth's surface. As it rises, gas bubbles form inside the magma. Finally, some of the magma breaks through weak areas in Earth's crust, such as cracks and **fissures**, and erupts. Once magma erupts it is called lava.

Types of Eruptions

Volcanic eruptions are sometimes very violent and fast. Other times, they are subtle and slow. The type of eruption depends on the magma. If the magma is thin and runny, it flows down the volcano as lava. Lava flows are less dangerous because they move slowly, giving people time to prepare and get out of the area.

Thick Magma

When magma is thick, sticky gas bubbles cannot escape easily. Pressure builds inside the magma as it makes its way to the surface. The result is an explosive eruption as the magma bursts through the vent. Violent explosions can be very dangerous. They can cause fast-moving, destructive fire clouds and blanket the area with ash.

A second blast from Mount St. Helens sent ash more than **12 MILES** (19 km) into the sky. Ash settled over several U.S. states and parts of Canada. It was the **LARGEST** and most **DESTRUCTIVE** volcano in U.S. history.

Kilauea in the Hawaiian Islands is the most active volcano in the world. It erupted with lava flows nearly nonstop from 1983 to 2018.

The Mount St. Helens eruption in Washington state on May 18, 1980, is an example of a violent explosion. A glowing cloud of gas and rock spewed from the mountain. It instantly destroyed everything within an 8-mile (13 km) radius.

Mountains vs. Volcanoes

Many people think volcanoes are mountains, but they are not. Mountains form when tectonic plates bump into one another and Earth's crust thrusts upward from the surface, folds inward, erodes, or **faults**. Volcanoes, however, form when flows harden and build up around the vent to form a mound. The vent connects through cracks in the volcano to a magma storage chamber deep inside the volcano. The volcano will continue to erupt in the same place each time so long as the vent remains connected to a source of fresh magma. As a result, volcanoes get bigger each time they erupt. Over time, a volcano may get so big that it looks like a mountain.

Types of Volcanoes

Volcanoes come in many different shapes and sizes, but there are four main kinds of volcanoes. Stratovolcanoes have steep, even sides and are made up of many layers of lava, ash, and rock. Many of the tallest mountains in the world are actually stratovolcanoes. Some stand more than 8,000 feet (2,438 m) above **sea level**. Lava dome volcanoes happen when lava is too thick to flow away from the vent. Instead, the lava piles up, over, and around the vent to form a mound with steep sides.

Japanese volcanic island Nishinoshima Shinto in an aerial photo from 2016. Located 584 miles (940 km) southeast of Tokyo, this island first formed with an eruption of an undersea volcano in 1702. It grew after further eruptions in 1973, and 2013–2015.

Mount Etna, on the east coast of the island of Sicily, in Italy, is a stratovolcano. It is the highest active volcano in Europe.

Shield Volcanoes

Shield volcanoes have long, gently sloping sides that take on the shape of a dome and look like a shield from a distance. They consist mainly of fluid lava. The lava comes out in all directions from a bowl-shaped vent at the summit of the volcano or from several cracks that spiral out from the summit. Over time, thousands of lava flows pile up on top of each other and spread over large areas. The largest volcanoes on Earth are shield volcanoes.

Cinder Cones

Cinder cone volcanoes form when small pieces of lava blast into the air, cool, and fall down to form a cone around the vent. Most cinder cones are round or oval in shape. They can rise as much as 1,000 feet (305 m) above the surrounding landscape. If there are many eruptions from the same vent, cinder cones can eventually become stratovolcanoes.

Craters and Calderas

A crater is a bowl-shaped dip in the ground that is created by volcanic activity. Craters form when rocks and other materials explode outward from the vent and pile up around it. Volcanic craters can be deep or shallow, narrow or wide. Most are less than 0.6 miles (1 km) in diameter. Calderas are similar to craters, but they are much larger, between 0.6 and 31 miles (1 and 50 km) in diameter. They form after a large eruption that empties out the magma chamber. The volcanic material above the empty chamber collapses inward to form a large **depression**.

Iceland is home to many shield volcanoes.

Cinder cone, Meke Crater Lake, Turkey

Volcanic crater, Kamchatka, Russia

Thermal Activity

A fumarole is a type of vent that releases gas and steam from volcanic activity below Earth's surface. Fumaroles are often seen between eruptions on active volcanoes. If there is an underground water source nearby, fumaroles can form hot springs. Hot springs typically occur in places where cool groundwater is heated by magma or hot rock. A geyser forms when steam pressure builds up inside a hot spring. When the pressure gets too much, the steam erupts and sends hot water gushing into the air.

Old Faithful is a famous cone geyser located in Yellowstone National Park in Wyoming. It spews water at a temperature of 204 degrees Fahrenheit (95.6 °C) about 17 times a day.

What Is Tephra?

Rock fragments that blast out of a volcano during an eruption are called tephra. Bombs or blocks are the largest form of tephra. They are bigger than 2.5 inches (64 mm) in size.

VOLCANIC ASH is the smallest form of tephra. It consists of rock fragments that are less than **0.08 TO 0.16 INCHES** (2 to 4 mm) in size.

Ash is one of the most common hazards to come from a volcanic eruption. Due to its small size, ash is easily carried great distances via wind and rain. It can damage water supplies, utilities networks, transportation systems, and buildings.

Lava Flows

Lava flows are the rivers of hot, molten rock that stream from an erupting volcano. The shape, width, length, and thickness of the flow depend on the type of lava, the slope of the volcano, and other factors. There are a few main types of lava flows. Pahoehoe flows have a smooth, shiny appearance and are silver in color. They are thinner and move more quickly because they have a runnier texture. A'a flows are slower and thicker. They contain rubble made from broken lava rocks and are dark gray in color. Blocky flows contain larger blocks of rubble and move even more slowly than a'a flows.

Pahoehoe lava

Blocky lava

A'a lava

Do All Volcanoes Erupt?

Not all volcanoes erupt. Volcanoes are put into one of three categories based on their likelihood to erupt. Active volcanoes have erupted in the past 10,000 years and have the potential to erupt again. They may even be erupting or showing signs of volcanic activity right now. **Dormant** volcanoes have not erupted in the past 10,000 years but have the potential to do so again at some point. **Extinct** volcanoes have not erupted in thousands of years, and there is no chance they will ever erupt again.

Mount Kilimanjaro in Tanzania, Africa, is an example of an extinct volcano.

The columns at the Giant's Causeway in Ireland are the result of magma that came up from fissures in rock to form a lava plateau 60 million years ago.

CASE STUDY

Volcanoes Triggering Volcanoes

For years, scientists have questioned if it is possible that a volcanic eruption could trigger another eruption at a nearby volcano. The answer is maybe. Some volcanoes are part of a complex network of cinder cones, stratovolcanoes, fissures, vents, and calderas that are all part of one large volcano. In other words, the same volcano is erupting in different places at roughly the same time.

Eruptions at Rabaul Volcano in Papua New Guinea are a good example of **simultaneous** volcano activity. Two vents located within the volcano, Vulcan and Tavurvur, erupted at the same time in 1937. The eruption killed more than 500 people. However, one eruption did not trigger the other. Magma from the same complex system simply rose to the surface in two places at once. This event led to the founding of the Rabaul Volcano Observatory, which monitors volcanic activity in the area. When the two vents erupted again in 1994 only five lives were lost. Thousands of others were saved due to careful planning.

CHAPTER 2

Studying Disasters and their Effects

Volcanology is the scientific study of volcanoes. Until the 1800s, the forces underlying volcanoes were largely unknown. Today, scientists study past volcanic events in order to help them understand how, when, and why volcanoes erupt. The more they know about these natural events, the more they can do to help prevent or reduce the impact of natural disasters.

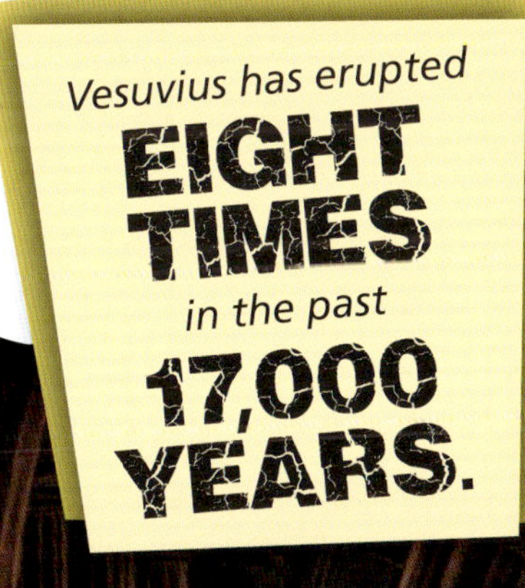

Vesuvius Erupts

Mount Vesuvius is a stratovolcano in Italy. It is one of the most studied volcanoes in the world. Vesuvius is best known for the massive eruption that took place in 79 C.E. The eruption was documented by Pliny the Younger. Pliny was a well-known lawyer and government official who wrote detailed letters about life in Rome. He was an eyewitness to the Vesuvius eruption when he was 17 years old. He wrote two detailed letters about the event 25 years after it happened. These letters are the first known volcanology reports. They have helped scientists piece together the different stages of the Vesuvius eruption. They have given historians insight into the thoughts, actions, and feelings of people at the time.

The eruption of Vesuvius happened in stages. The first blast sent gases, rocks, and a cloud of pumice into the air. Although they had time to escape, many people in Pompeii remained. Two days of eruptions brought more ash and a surge of superheated gases.

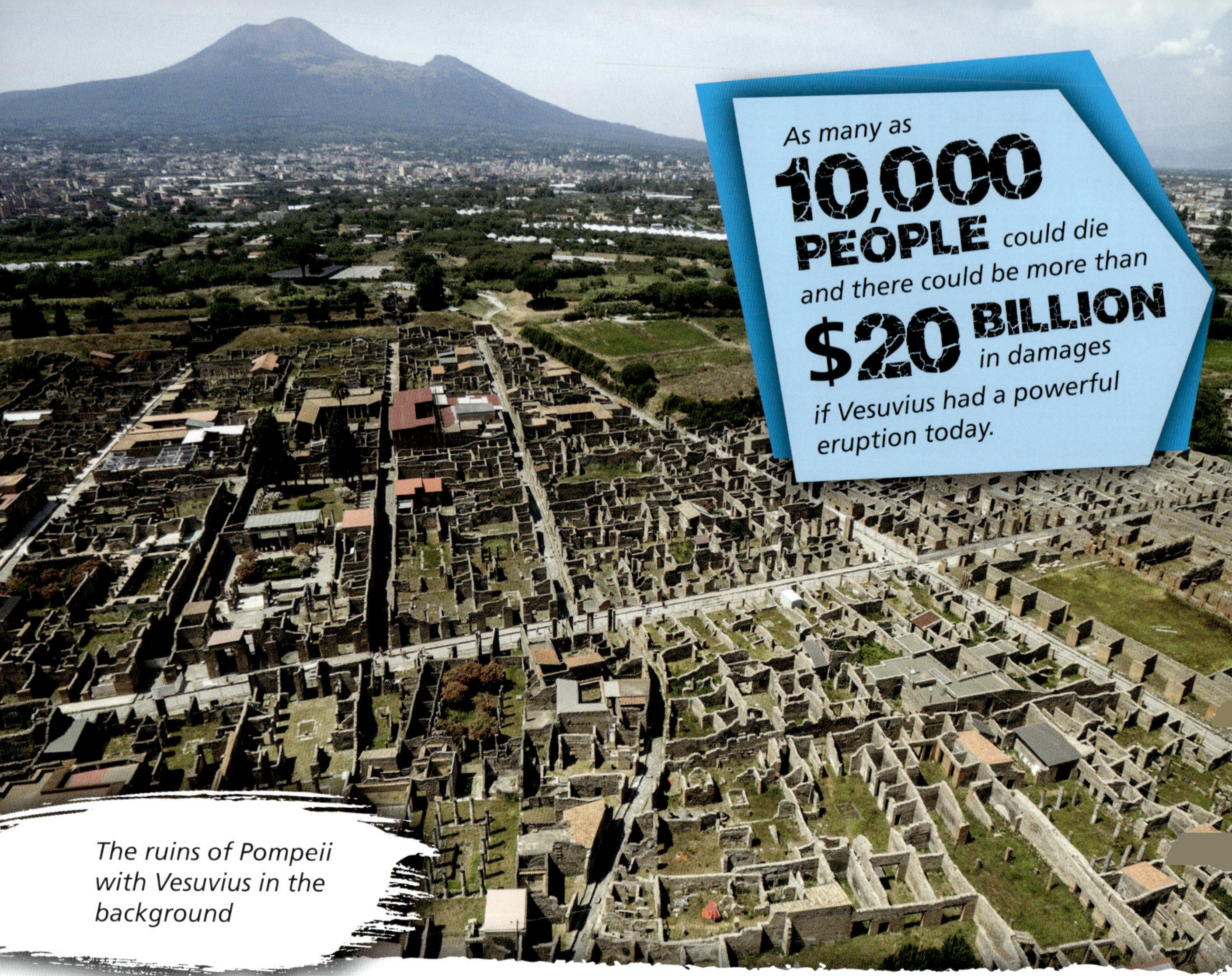

The ruins of Pompeii with Vesuvius in the background

Pompeii vs. Vesuvius

Residents in the nearby community of Pompeii did not know Vesuvius was going to erupt. They did not prepare or evacuate. Fast-moving **pyroclastic flows** rushed down the slopes of the volcano and destroyed everything in their path. Meanwhile, huge plumes of ash and gas exploded miles (km) into the sky. Nearly everyone in the area was killed and entire towns were left devastated. In the 16 years leading to the eruption, there were a series of earthquakes in the area. Without modern science, residents had no way of knowing about the connection between seismic activity and volcanoes.

Studying the Disaster

Scientists have learned a lot about the eruption from perfectly preserved ruins and human remains in Pompeii and other towns that were destroyed by the volcano. In one study, scientists analyzed the structure of rock and ash deposits from the eruption. They were able to map the speed, temperature, path, and other details of the pyroclastic flows that swept through Pompeii. The study helped scientists better understand how pyroclastic flows behave so that they can help people similar natural events in the future.

Vesuvius Today

Vesuvius remains one of the most dangerous volcanoes on Earth. Scientists predict it could erupt at any time. About 700,000 people live within its danger zone. Another 3 million people live in the nearby city of Naples, Italy. If Vesuvius were to erupt, they would be in danger of pyroclastic flows. The heat would be so intense that it could consume a human in one second.

There are many ongoing studies on Vesuvius and plans in place to reduce the impact of an eruption. In addition, the Vesuvius Observatory continuously watches for seismic activity in the area. The local government has created an evacuation plan that would quickly move hundreds of thousands of people out of the danger zone within 72 hours of an eruption.

SCIENCE BIO

Massimo Orazi, Vesuvius Observatory

Mount Vesuvius monitoring station

Massimo Orazi works with the seismic monitoring unit at the Vesuvius Observatory. Established in 1841, the Vesuvius Observatory monitors the volcanic regions of Campania, Mount Vesuvius, the Phlegraean Fields supervolcano near Naples, and the volcanic island of Ischia in the Gulf of Naples. It is part of Italy's National Institute of Geophysics and Volcanology and is the oldest scientific institution dedicated to the study of volcanoes.

Orazi's seismic monitoring unit consists of a network of stations near Vesuvius that collects data from seismologic sensors and sends them back to a data center in **real-time**. Orazi develops all the electronics needed for the network to function, from the sensors to the data center. His seismic monitoring equipment is used in the field and the lab with the direct goal of protecting nearby populations through early warning.

Waterfall Effect

Scientists are also interested in protecting against other natural hazards triggered by volcanic activity, such as landslides and tsunamis. On December 22, 2018, a massive tsunami struck the Indonesian Islands of Sumatra and Java. Local residents were completely unprepared for the natural event, which resulted in a major natural disaster. Hundreds of people died and thousands more were injured.

Anak Krakatau Erupts

The tsunami was triggered by the collapse of a nearby offshore volcano, Anak Krakatau. The volcano had been erupting lava and ash since June 2018.

Early Warning Systems

Indonesia has an early warning system to alert locals of earthquakes that could cause tsunamis. However, the system does not detect landslides or volcanoes, even though landslides also caused a tsunami in Maumere in 1992 and in Palu in 2018. Indonesia's tsunami buoy system has not been operational in the area since 2012.

Anak Krakatau, Indonesia

Indonesia's buoy system consists of 21 buoys that use deep-sea sensors to gather data and send out advance warnings of massive waves. It was put in place in 2004 after a tsunami ripped through the country and killed almost 250,000 people. Since then, some of the buoys have been stolen or damaged. The government does not have the money to fix or replace the system.

Satellite images showed a **158-ACRE** *(64 h) portion of Anak Krakataua sliding off into the ocean. The result was an underwater landslide that created massive waves nearly* **10 FEET** *(3 m) tall.*

Land of Ice and Snow

Sometimes, even if no one dies and there is not a lot of damage following a volcanic eruption, there is still a huge price to pay. Volcanoes can have long-term effects on **economies** and the environment. This was the case when the Eyjafjallajökull volcano erupted in Iceland. Eyjafjallajökull lies beneath a glacier just 99 miles (160 km) from the nation's capital, Reykjavik. The eruption began in March 2010 and lasted seven months. While the volcano has erupted six times in history, this was the first time since the 1820s.

Staying Safe

The Icelandic Meteorological Office monitors Eyjafjallajökull regularly so the government can alert locals of any volcanic activity. When residents in any country have advanced notice of a potential eruption, a natural disaster can be prevented, as was the case in Iceland in 2010. Eyjafjallajökull is located on a massive ice sheet. Heat from the eruptions caused floods as the ice melted. Flooding washed out roads and bridges in some places. Gases released by the eruption killed vegetation, and ash **contaminated** water supplies.

Ongoing Studies

The ash from Eyjafjallajökull was very fine, which posed a problem for airlines. Ash can clog airplane engines, so many flights from North America to Europe were cancelled or redirected to avoid the ash. Following a major eruption in April 2010, more than 300 airports in about two dozen European countries closed for nearly a week. More than 100,000 flights were cancelled. The Eyjafjallajökull eruption led to major studies on airplane engines all over the world. Scientists did not know much about the effect of ash on aircraft. They began to look for ways to keep planes in the sky even if there is ash in their path. Some scientists studied how much ash airplanes can handle. Others tested new technologies, such as ash-proof engines and ash-**resistant** coatings for engines.

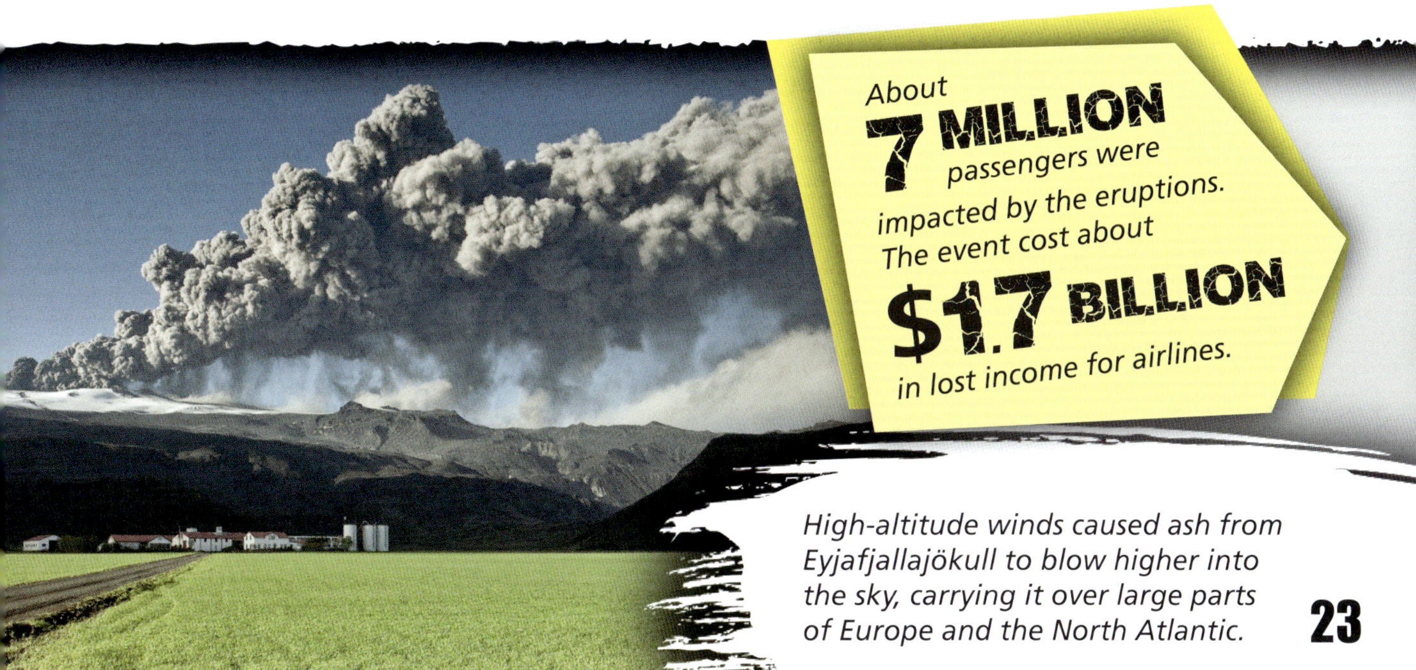

About **7 MILLION** passengers were impacted by the eruptions. The event cost about **$1.7 BILLION** in lost income for airlines.

High-altitude winds caused ash from Eyjafjallajökull to blow higher into the sky, carrying it over large parts of Europe and the North Atlantic.

CHAPTER 3

Meeting the Challenge

Early detection of eruptions is key to reducing the risks volcanoes pose to humans and property. Volcanologists use special techniques and tools to tell if a volcano will reawaken after being dormant for a long time. They study conditions at volcano sites and research past events to get a sense of a volcano's historic behavior.

Working in the Field

Between eruptions, volcanologists visit volcanoes to observe the landscape. They look for steaming vents, new cracks in the ground, and changes to mineral deposits or plant life. Volcanologists often take special steps to prepare for working in the field. They may take safety training, learn first aid, and practice their wilderness **survival skills**. Once on site, volcanologists may need to wear gas masks and heat-resistant clothing to protect themselves from the **elements**.

Volcano Monitoring

Volcanologists keep a diary of their findings in the field. They do not touch anything on the volcano until they have made sketches and notes. They write about the different rock layers and describe what the volcano looks like. They take pictures and keep records in their computers.

Volcanology is a dangerous job. When working in the field, volcanologists risk their lives. Hazards include gas emissions, steam explosions, earthquakes, and eruptions.

A volcanologist samples lava at the Kilauea volcano in Hawaii.

A volcanic bomb from Vesuvius filled with the mineral olivine and other crystals.

Rock Samples

Volcanologists collect rock samples to take back to the lab to investigate. They use a rock hammer to chip off pieces of **igneous** rock that forms from hardened magma. They also use hollow steel pipes to drill into the ground and collect core samples of earth and rock. They use the samples to date past volcanic events.

Scientists use different techniques back in the lab to study the rocks they find in the field. One, called an electron microprobe, measures the types of crystals and minerals found inside small rock fragments. Knowing what the rocks are made from helps scientists figure out how they formed and how old they are.

Surveying the Ground

Scientists look for changes in the size and shape of a volcano. This is called ground **deformation** monitoring. Changes on the outside of the volcano are often due to the movement of magma inside the volcano. Any cracking, sinking, or swelling of the ground offers clues about when an eruption might take place. Volcanologists use special technologies to collect real-time data from inside and outside of volcanoes. The data is then sent to observatories to analyze.

Tools of the Trade

One of the most common tools volcanologists use is the **Global Positioning System (GPS)**. GPS receivers are set up around the volcano. The receivers detect and record any changes in the shape of the volcano. The data is sent to **satellites** that orbit Earth from space. Volcanologists look at all of the data that comes from the receivers over a certain period of time to see if the ground has moved.

The volcano's crust may start to shift and bulge as magma rises to the surface. Volcanologists use a device called a seismometer to measure ground vibrations. It helps them track magma as it moves up and down inside the volcano. Volcanologists also use spirit-levels, electronic tiltmeters, and electronic-laser beam instruments to measure changes in the shape of a volcano. Pressure builds up inside a volcano just before it erupts, causing it to swell. These devices can detect small changes.

A GPS receiver monitors the Newberry volcano in Oregon.

GPS *is very precise. It can detect changes as small as* **0.4 INCHES** *(1 mm) per year.*

This radar system on the rim of the Halema'uma'u at Kilauea in Hawaii uses microwave pulses to measure the height of the lava lake.

SCIENCE BIO
David A. Johnston

Dr. David A. Johnston was a volcanologist who specialized in gas sampling. He joined the U.S. National Science Foundation in 1978 and spent a lot of time working in the Pacific Northwest. Johnston was the first volcanologist to arrive at Mount St. Helens in 1980 when the volcano began to show signs of activity after more than 100 years. He took gas samples from the volcano both in the air and on the ground as part of a study for the U.S. Geological Survey (USGS).

Johnston was stationed just 5 miles (8 km) from the volcano when it erupted on the morning of May 18, 1980. The last words he spoke before his death were to warn others of the eruption. He said into his radio transmitter, "Vancouver! Vancouver! This is it!" Two years later, the USGS dedicated the David A. Johnston Cascades Volcano Observatory (CVO) in Vancouver, Washington, in memory of the volcanologist who gave his life to his studies.

David A. Johnston, working on Mount St. Helens hours before his death.

Volcanic Explosive Index

Volcanologists use a scale called the Volcanic Explosive Index (VEI) to compare the explosiveness of volcanic eruptions. They measure the volume of matter that erupts from the volcano. They also measure how high it shoots into the air and other factors, such as how often the volcano erupts. The eruption is then assigned a value based on the VEI.

The VEI is not an ideal scale for measuring and tracking volcanic explosions. It only measures the matter ejected from the volcanic eruption rather than the power behind it. Scientists are trying to create a new system that also measures the **amplitude** of the eruption.

VEI	Tephra volume	Frequency	Example
0 *nonexplosive*	*35,314 ft³ (1,000 m³)*	*daily*	*Dallol 2011*
1 *gentle*	*353,146 ft³ (10,000 m³)*	*daily*	*Raoul Island 2006*
2 *explosive*	*35,314,666 ft³ (1,000,000 m³)*	*weekly*	*Whakaari 2019*
3 *severe*	*353,146,667 ft³ (10,000,000 m³)*	*yearly*	*Ontake 2014*
4 *cataclysmic*	*3,531,466,672 ft³ (100,000,000 m³)*	*10 years*	*Calbuco 2015*
5 *paroxysmal*	*0.24 miles³ (1 km³)*	*100 years*	*Mount Saint Helens 1980*
6 *colossal*	*2.4 miles³ (10 km³)*	*100 years*	*Pinatubo 1991*
7 *supercolossal*	*24 miles³ (100 km³)*	*1,000 years*	*Tambora 1815*
8 *megacolossal*	*239.9 miles³ (1,000 km³)*	*10,000 years*	*Yellowstone 630,000 BCE*

The largest volcanic explosions in history are given a value of eight on the VEI.

Monitoring Volcanic Gases

Water vapor, carbon dioxide, and sulfur dioxide account for 99 percent of the gases that erupt from a volcano. Scientists study the levels of these gases coming from the vent of active or potentially active volcanoes to help predict eruptions. Gases can tell scientists a lot about what is going on inside a volcano. The composition of the gases change as magma rises to Earth's surface. One way to monitor these changes is to mount special devices on aircraft. The devices collect data on how much gas is let off by the volcano each day. Devices that detect carbon dioxide can also be placed directly on a volcano to collect data. Satellites in space are equipped with devices that can measure sulfur dioxide in volcanic clouds.

Volcanologists collect samples of rocks at a volcano fumarole in Russia.

Spectrometers

Volcanologists at the USGS Hawaiian Volcano Observatory (HVO) study gases that come out of Kilauea Volcano. They use a device called a **spectrometer** to measure the volcano's sulfur dioxide **emissions**. Small spectrometers are attached to a vehicle and then driven under the volcano's plume. They measure the amount of ultraviolet (UV) radiation energy absorbed by the volcano's gas plume as sunlight passes through it. Volcanologists at Kilauea also use small infrared analyzers to measure carbon dioxide levels. Like the UV spectrometers, they attach the analyzers to vehicles and drive them through the plume.

Hawaii's Kilauea volcano caldera is shown here at night.

A United States Geological Survey geologist uses a UV spectrometer to detect gases at a fumarole on Mount Hood, Oregon.

Studying the Past

To learn about volcanic events that happened in the past, scientists must compare the rocks from the ancient past with rocks from today. They use a technique called radiocarbon dating to find the age of fossils found in rocks, core samples, and sediment. Every living thing, including leaves, wood, and animals, contains a certain amount of **carbon**. The carbon begins to decay when the organism dies. The older the fossil is, the less carbon it contains. Scientists can find the age of fossils that are up to 50,000 years old using radiocarbon dating.

Dating Rocks

Potassium-argon dating is a method scientists use to learn the age of volcanic rocks. Argon is a gas found in rocks. When a rock melts, the argon escapes. When the rock cools, all that is left is potassium. Over time, the potassium decays and the amount of argon in the rock increases. Scientists compare the **ratio** of argon to the ratio of potassium to find out how old the rock is.

The oldest rocks on Earth are more than **4.5 BILLION** *years old.*

Raufarhólshellir is the fourth-longest lava tube in Iceland. Based on carbon dating we know it formed 5,600 years ago. Lava tubes are tubes that drain lava during an eruption. The tubes form caves of hard rock when the lava flow ceases.

Advance Warning

Even if a volcano is not active, it is important to continually monitor its behavior. A volcano can erupt at any time without notice. Scientists cannot predict the exact time or date of an eruption. They can, however, provide real-time updates about changes in volcanic activity to communities at risk. If it seems like a volcano might erupt soon, volcanologists send out eruption warnings. They let people know of the danger so they can prepare and leave the area if needed.

A map shows a real-time update of the amount of tephra from eruptions of volcanoes in the Cascade Range.

One of the best ways to help prevent a natural disaster is to educate people about the dangers and how to react to warnings. Volcanic hazards maps are used to show people who live in high-risk areas the specific types of hazards they might face. This gives people the information they need to prepare for a natural event.

Lahars

Lahars are mudflows that mix with pyroclastic flows. They are common in places where there are rivers, heavy rainfall, or large amounts of ice and snow. Lahars move very quickly and cause major damage. They can move boulders, cars, and buildings. When they stop moving, they become solid, much like cement.

Advance Warning Systems

In places where there are snow-covered volcanoes, such as Mount Rainier, lahar detection systems are used. These systems help provide advanced warning so that people can evacuate the area. Such a system would have been helpful when the Nevado del Ruiz volcano in Colombia erupted in 1985.

The eruption was not large, but it produced enough heat to melt ice and snow from a glacier at the top of the volcano. Ice water mixed with pyroclastic flows to create three lahars. The lahars raced down the side of the volcano at speeds of up to 19 miles per hour (30 kph). They destroyed everything in their path. More than 23,000 people were killed by the lahars in just a few hours.

The Nevado del Ruiz eruption was Colombia's worst natural disaster, the the second-deadliest volcanic disaster of the 1900s, and the fourth-deadliest in recorded history.

The tragic event led to the creation of the international Volcano Disaster Assistance Program (VDAP), which monitors all of the active volcanoes on Earth, as shown on this map.

At 62 miles (100 km) away, the city of Tokyo, Japan is within the "shadow" of Mount Fuji. Climbing the still-active volcano is a popular activity for locals and tourists.

CASE STUDY

Mount Fuji Disaster Management

Japan's Mount Fuji has been dormant since 1707. But scientists warn that it could erupt again soon. This would place more than 750,000 people in danger. In 2018, the Japanese government's Central Disaster Management Council launched a major study into Mount Fuji. They found that an eruption the same size as the one that took place more than 300 years ago could cover central Tokyo with about 0.4 to 0.6 inches (1 to 1.5 cm) of ash over 15 days. Under the right conditions, the ash could cut off power supplies and prevent aircraft from flying in the area. It could also pile up so high in some places that roads would become blocked and houses could collapse. Based on these findings, the council is looking for ways to prevent a natural disaster in the event of an eruption. Some of the areas they hope to address include traffic control, ash disposal, and the safe evacuation of people from the area.

CHAPTER 4

Facing Future Disasters

Volcanic eruptions are as old as the Earth. There is no successful way to prevent them from happening. Unlike earthquakes, floods, and hurricanes, there are no special methods for constructing volcano-proof buildings that can withstand hot lava or prevent ash from falling on them.

Preparation is Key

Scientists are always investigating new technologies to predict and prevent volcanic eruptions. For now, though, the best way to prevent natural disasters due to volcanoes is to prepare for them. People living in volcanic danger zones need to get educated about the hazards and take the right precautions.

Climate Change

Every 20 years or so, there is a volcanic eruption that is so big it causes a small, short-term change to Earth's climate. The gas and other materials that spew from these massive eruptions shield the Sun's rays from reaching Earth. The result is a period of global cooling that lasts about two years.

Global Warming?

Some people believe natural events, such as volcanic eruptions, contribute to **global warming**. However, eruptions account for only a small percentage of the carbon dioxide released into Earth's atmosphere. Human activities cause much more harm than volcanoes. In fact, some scientists believe volcanic eruptions may start to happen more often due to global warming.

Scientists have found a link between melting glaciers, rising sea levels, and volcanic eruptions. They studied past events and found that as glaciers melt, there is less pressure on Earth's crust. Magma can rise to the surface more easily. As a result, it stands to reason that as the global climate increases and more glaciers melt, there will be more volcanic activity.

Fracking Impact

Hydraulic fracturing, or fracking, is a method used to drill into deep underground rocks. The purpose of fracking is to extract natural gas from inside the rocks. As the drill sinks into the ground, it creates small cracks and fissures in the rocks. A mixture of water, sand, and chemicals is then blasted at the cracks to open them wider. Natural gas then flows from the cracks to a well at the surface. Many people worry that fracking could cause earthquakes or volcanic eruptions. In fact, there have been reports that fracking caused volcanic activity in the Puna region of Hawaii's Big Island in May 2018. However, scientists have not found any evidence that this was the case. In fact, there were no fracking companies working in the area at the time and the eruption first started in 1983.

Volcanoes are a great source of geothermal energy. Turbines can be used to collect underground sources of steam and water that are heated by magma. Geothermal energy can be used as a source of electricity.

Living Near a Volcano

Given the dangers, why would anyone live near a volcano? Some people simply cannot afford to move. Others do not want to leave their friends and family behind. Despite the dangers they pose, volcanoes also bring many benefits to an area. Once dry, lava contains valuable minerals, such as gold, zinc, copper, and diamonds, that can be mined. It also produces nutrient-rich fertile soils that are perfect for farming.

One of the world's largest deposits of potash is located near the Dallol volcano in Ethiopia. It is not mined commercially.

Volcano Tourism

Millions of tourists visit volcanoes each year. Many choose to hike or bike along volcanic slopes. Others prefer to view volcanoes from a distance. Tourism creates many jobs for locals, but it can still be very dangerous if the volcano is active. Thirty-one people, including tourists and guides, were killed when a New Zealand island volcano erupted on December 9, 2019. Many others were severely burned when the Whakaari Island volcano erupted, releasing steam, gases, and rocks into the air. The volcano had been showing signs of activity for weeks. A former president of the International Association of Volcanology and Chemistry of the Earth's Interior said that the eruption was "a disaster waiting to happen" and people should not have been allowed to visit.

Some people live near volcanoes for cultural or religious reasons. Mount Paektu in North Korea and Kilauea in Hawaii are examples of sacred volcanoes that could blow at any time. However, locals believe the sacrifice is worth it to live on a beautiful volcano that has cultural or religious importance.

When measured from its underwater base, Mauna Kea in Hawaii is the tallest mountain in the world. It is a sacred site to the Hawaiian people. The dormant volcano is now occupied by many large scientific telescopes.

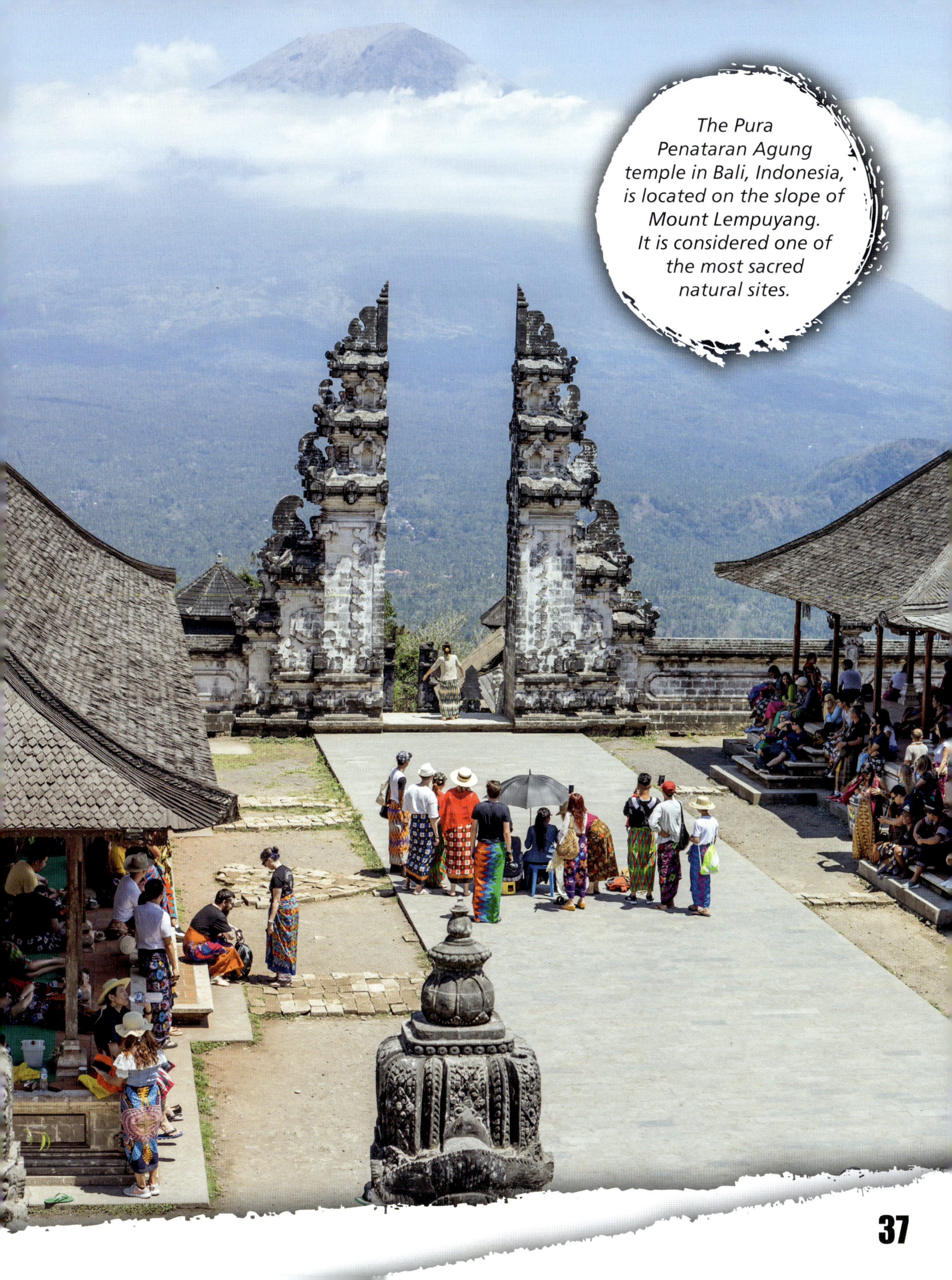

The Pura Penataran Agung temple in Bali, Indonesia, is located on the slope of Mount Lempuyang. It is considered one of the most sacred natural sites.

Be Prepared

People who live in volcanic regions should always be prepared for an eruption. They should be aware of the hazards and have a disaster plan in place. They need to know their evacuation routes and where to seek shelter. It is also a good idea to have an emergency kit on hand. The kit should include first-aid items, canned foods, blankets, water, dust masks, a flashlight, and a battery-powered radio.

Stay Aware

During an eruption, it is important to listen to the radio for any instructions from local authorities. Keep doors and windows closed to prevent ash and debris from coming in. Move cars, farming equipment, and other machines indoors or cover them with tarps. Make sure all animals are sheltered to keep them safe as well.

After an eruption, avoid going outdoors, especially if there is heavy ashfall. People should keep their nose and mouth covered to protect their lungs. It is also a good idea to wear long sleeves and pants, as well as goggles. Once safe, homeowners should clear any ash and debris off the roof of their house so it does not cave in under the weight.

A layer of dry volcanic ash **4 INCHES** (10 cm) thick weighs **120 TO 200 POUNDS** (54.4 to 90.7 kg).

NORMAL	*Volcano is in typical background, noneruptive state or, after a change from a higher level, volcanic activity has ceased and volcano has returned to noneruptive background state.*
ADVISORY	*Volcano is exhibiting signs of elevated unrest above known background level or, after a change from a higher level, volcanic activity has decreased significantly but continues to be closely monitored for possible renewed increase.*
WATCH	*Volcano is exhibiting heightened or escalating unrest with increased potential of eruption, timeframe uncertain, or eruption is underway but poses limited hazards.*
WARNING	*Hazardous eruption is imminent, underway, or suspected.*

The USGS's volcano alert chart rates the severity of volcanic eruptions.

When the volcano alert level is changed, a Volcano Activity Notice (VAN) is issued.

Volcano Warning Systems

In most cases, local governments are responsible for monitoring volcanic activity. They work closely with volcano experts to alert the public about potential dangers. There are various levels of alerts depending on the situation. In the United States, the USGS created a system of alert levels and color codes to describe the conditions at a volcano site. Scientists working at the five Volcano Observatories across the country use this system to send out volcano alerts. Other countries, such as Russia, Colombia, Indonesia, and New Zealand, have similar systems.

Emergency Response

The Red Cross, Oxfam International, and other **humanitarian** organizations provide aid to people impacted by eruptions. These organizations rush to communities that are hit the hardest to help people in need. They set up emergency shelters for people whose homes have been destroyed. They bring food, clean drinking water, bedding, and other supplies to the area. They also provide medical care to people who have been injured.

Slowing the Flow

One way to prevent damage from fast-moving lava flows is to slow them down. In 1973, Eldfell volcano erupted on the island of Heimaey in Iceland. Most people were safely evacuated within 24 hours of the eruption. However, the eruption lasted for about six months. It destroyed hundreds of homes and covered most of the area in a thick layer of ash. Lava flows threatened to destroy the harbor, which would have a major impact on the local economy. Locals were desperate to save it. They sprayed more than 8 million cubic yards (6 million m^3) of seawater onto the lava to cool it down and slow the flow.

Reducing the Pressure

When there is too much pressure inside a volcano, tiny gas bubbles form in the magma. As the pressure decreases, the bubbles escape out the top of the volcano. Magma comes spewing out as well. In theory, reducing the pressure inside a volcano could help stop an eruption. Scientists think it may be possible to release some of the gases by drilling into magma chambers. This would help **depressurize** volcanoes and possibly prevent an eruption.

Triggering Eruptions

Scientists have thought about using explosives to trigger an eruption for many decades. During World War II, a volcano expert suggested dropping bombs into volcano vents in Japan. He wanted to set off an eruption of ash and lava. The plan was never put in place. However, the idea of setting off an eruption on purpose could help reduce the impact. By planning when an eruption takes place, people could be moved safely out of the way in advance. For instance, Mount Rainier is a volcano that stands very near Seattle, Washington, which has a population of more than 700,000. An eruption on the northwest side of the volcano would be a major threat to the city. Purposely setting off an eruption on the east side of the volcano would cause much less damage as there are only a few small communities in that direction.

Lava Barriers

For centuries, people have been digging channels to direct lava flows away from communities. The first known attempt to control the flow of lava was in 1669. The people of Catania, Sicily, dug a trench to divert lava pouring out of Mount Etna. However, the lava then flowed toward a nearby town instead. In 1992, Mount Etna erupted again. Locals used a different method to halt lava flows. First, they built dams from dirt, but the lava poured over top of them. Later, helicopters dropped concrete blocks at the edge of the crater to plug the lava flow. The barrier lasted long enough to dig a trench that would channel the lava away from nearby communities.

The city of Seattle with Mount Rainier in the background.

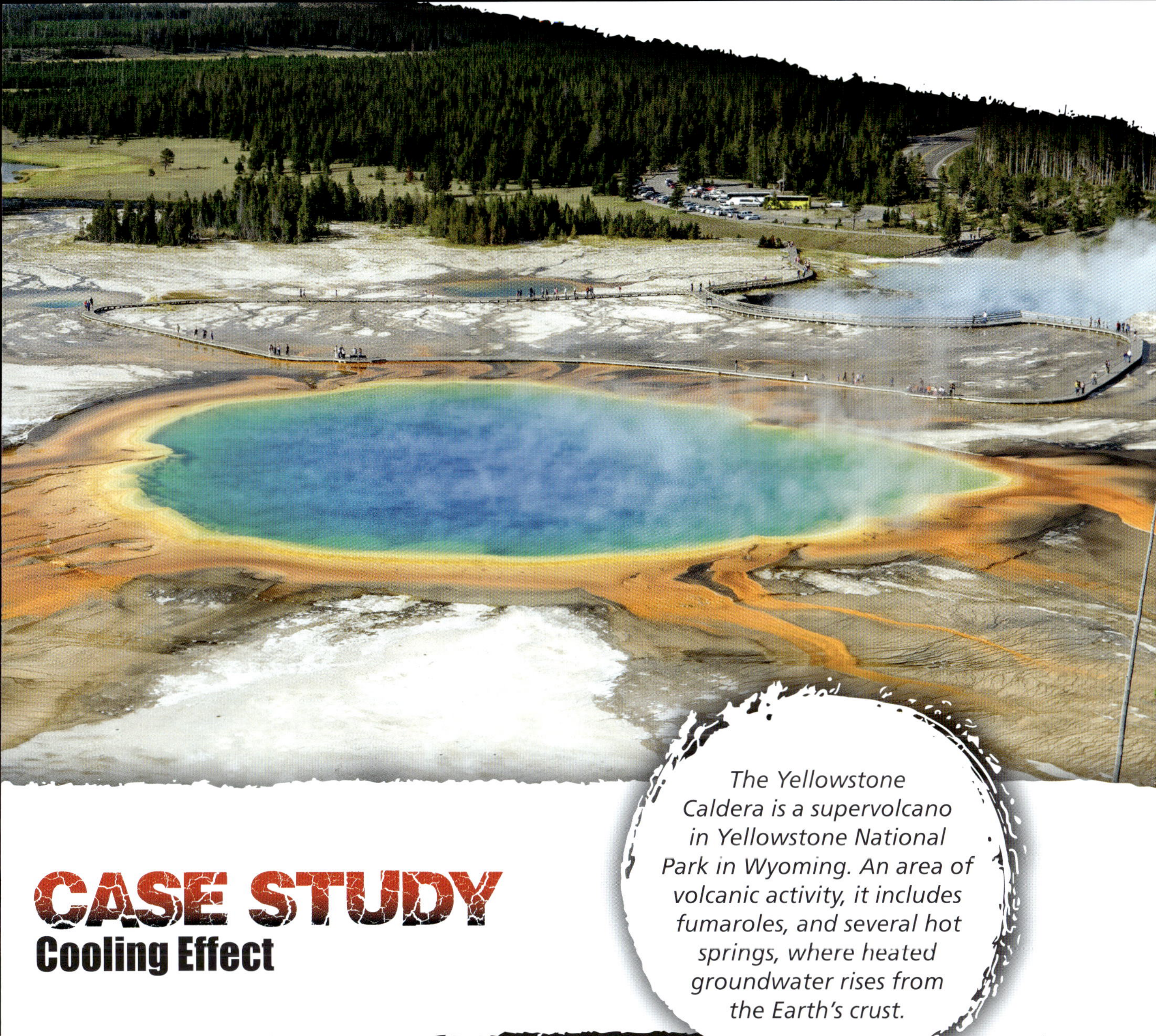

The Yellowstone Caldera is a supervolcano in Yellowstone National Park in Wyoming. An area of volcanic activity, it includes fumaroles, and several hot springs, where heated groundwater rises from the Earth's crust.

CASE STUDY

Cooling Effect

A massive supervolcano thousands of times more powerful than a normal volcano lies beneath Yellowstone National Park in the United States. An eruption would spread ash across the country and cause major damage. However, the Yellowstone supervolcano has only erupted three times in history. The most recent eruption was more than 660,000 years ago. The chance of it blowing in the near future is slim.

Still, volcanologists think about how they might stop an eruption if needed. For instance, NASA is researching one **innovative** idea. It would begin by drilling holes 5 miles (8 km) deep in the hot groundwater that surrounds the magma chamber. Then NASA would pump huge amounts of water through holes. The water would make the magma so thick and sticky that it would not be able to rise to the surface. The project would take hundreds of years to complete and cost more than $3.5 billion.

CHAPTER 5

Conclusion

Although volcanic eruptions are quite common, few become true, catastrophic disasters. This is because we are much better prepared for eruptions today than we were in the past. Scientists are always looking for new ways to predict when volcanoes might erupt so they can help people stay safe.

By studying the past and using new technologies, scientists have learned a lot about how volcanoes erupt and the signs to look for so that communities can take the necessary safety steps. Using advanced warning systems, they can alert people of the dangers they face and help them evacuate the area.

Over the past century, thousands of lives have been saved thanks to advances in our knowledge of volcanic eruptions. While we cannot stop these natural events from happening, we can try to reduce their impact on humans. The best way to prepare for an eruption is to educate people living in danger zones. When people know the risks they face and what to do before, during, and after a volcanic eruption, they can take an active role in minimizing the impact of natural disasters.

13,680 FEET

At 13,680 feet (4,170 m), Mauna Loa in Hawaii is the second largest volcano on Earth.

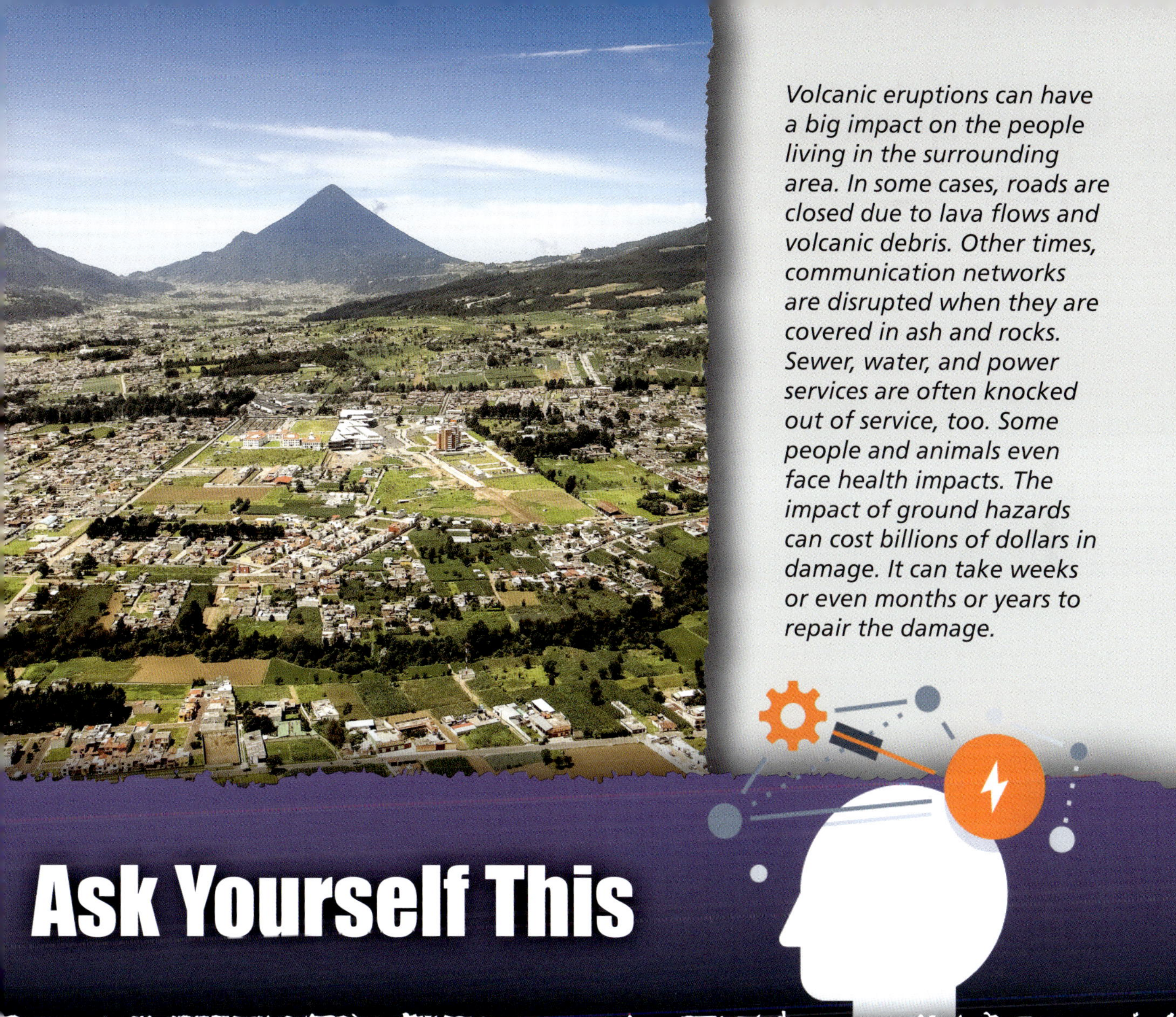

Volcanic eruptions can have a big impact on the people living in the surrounding area. In some cases, roads are closed due to lava flows and volcanic debris. Other times, communication networks are disrupted when they are covered in ash and rocks. Sewer, water, and power services are often knocked out of service, too. Some people and animals even face health impacts. The impact of ground hazards can cost billions of dollars in damage. It can take weeks or even months or years to repair the damage.

Ask Yourself This

Based on the information in this book, what are some of the things humans have learned from devastating volcanic eruptions and their aftereffects?

1. What can we learn from studying past volcanic eruptions? How can they help us prevent natural disasters in the future?

2. Imagine you live in a volcano danger zone. What can you do to prepare for a volcanic eruption? What signs can you watch for to tell if the volcano is about to erupt? How would you know if you need to evacuate your home? Does your community have any measures in place to reduce the impact of a natural disaster?

3. How do scientists predict when volcanoes might erupt? What types of technology do they use?

4. What techniques can people use to try to stop volcanic eruptions or lessen their impact?

Bibliography

Introduction

"10 Interesting Facts About Volcanoes." Universe Today. https://bit.ly/2NTGGV8

Bagley, Mary. "Volcano Facts and Types of Volcanoes." Live Science, February 7, 2018. https://bit.ly/2GkwReG

Dickinson, Greg and Hugh Morris. "Is volcanic activity on the rise because of climate change—and if so, where's next?" *The Telegraph*, December 10, 2019. https://bit.ly/2RJPMET

Redd, Nola Taylor. "Olympus Mons: Giant Mountain of Mars." Space, December 9, 2017. https://bit.ly/2NTGDZs

"Remembering Mount Pinatubo 25 Years Ago: Mitigating a Crisis." USGS, June 13, 2016. https://on.doi.gov/2RmGVKr

Rosenberg, Matt. "The Mount Pinatubo Eruption in the Philippines." ThoughtCo, January 20, 2019. https://bit.ly/2TStL9u

Wei-Haas, Maya. "Volcanoes, explained." *National Geographic*, January 15, 2018. https://on.natgeo.com/2sRWu3j

Zimmermann, Kim Ann. "Io: Facts about Jupiter's Volcanic Moon." Space, August 14, 2018. https://bit.ly/2NTxiR3

Chapter 1

"10 Interesting Facts About Volcanoes." Universe Today. https://bit.ly/30Su71l

Bagley, Mary. "Mount St. Helens Eruption: Facts & Information." Live Science, October 16, 2018. https://bit.ly/2uvqgep

"Footage of the 1980 Mount St. Helens eruption." *Air & Space Magazine*. https://bit.ly/2NTxmjL

Harris, Tom. "How Volcanoes Work." HowStuffWorks. https://bit.ly/2sVBUiy

"Hot Springs and Geysers." *Encyclopedia Britannica*. https://bit.ly/2Rpiaxf

"How Do Volcanoes Erupt?" USGS. https://on.doi.gov/2TQk1MZ

"How is a volcano defined as being active, dormant, or extinct?" Oregon State University. https://bit.ly/30QiaZX

Oskin, Becky. "What Is Plate Tectonics?" Live Science, December 19, 2017. https://bit.ly/2TQy1X7

Pariona, Amber. "Difference Between An Active, Dormant, And Extinct Volcano." WorldAtlas, September 19, 2019. https://bit.ly/3ayWxCd

"Paul Segall." Stanford University. https://stanford.io/2GjGj1O

"Six types of eruptions." *Encyclopedia Britannica*. https://bit.ly/2sQrssr

"The Big Question: Why do volcanoes erupt" BBC, May 13, 2018. https://www.bbc.co.uk/newsround/44100737

The Editors of Encyclopedia Britannica. "Kilauea." *Encyclopedia Britannica*, May 24, 2018. www.britannica.com/place/Kilauea

"What is a volcano?" USGS. https://on.doi.gov/30M0azY

Chapter 2

"Aviation." Volcanic Ashfall Impacts Working Group. https://on.doi.gov/30SufxR

"Eyjafjallajökull : Iceland volcanic ash cloud 2010." Explore Volcanoes. https://bit.ly/37kkgE8

Gurioli, Lucia. "Feverish Effort Under Way to Understand Mt. Vesuvius." National Science Foundation. https://bit.ly/2tBsusY

Haynes, Suyin. "A Tsunami Has Killed Hundreds in Indonesia. Here's What to Know About the Latest Deadly Natural Disaster to Hit the Country." *Time*, December 25, 2018. https://bit.ly/2sWX2VG

Martin, Lisa and Naaman Zhou. "Indonesia tsunami caused by collapse of volcano." *The Guardian*, December 24, 2018. https://bit.ly/36ii0f7

Morton, Mary Caperton. "Of airplanes and ash clouds: What we've learned since Eyjafjallajökull." Earth, April 2, 2017. https://bit.ly/2NSL3Qc

Pain, Elisabeth. "Monitoring the Pulse of the Mount Vesuvius." American Association for the Advancement for Science, August 12, 2005. https://bit.ly/2vcMmTn

Sims, Kelsey and Kaitlyn Lavoie. "Case Study of Mt Vesuvius." Sutori. https://bit.ly/2RNw37r

The Editors of Encyclopedia Britannica. "Pliny the Younger." *Encyclopedia Britannica*, November 1, 2019. https://bit.ly/2GkvZqq

University of Alaska Fairbanks. "UAF scientists collaborate to study Eyjafjallajokull lightning." Science X, May 25, 2010. https://bit.ly/3azqqC9

Wallace-Hadrill, Andrew. "Pompeii: Portents of Disaster." BBC, March 29, 2011. https://bbc.in/30Mj1et

"What if Vesuvius Erupted Today?" CBC, September 3, 2017. https://bit.ly/2TPmyXS

Wei-Haas, Maya. "Why Indonesia's 'volcano tsunami' gave little to no warning." *National Geographic*, December 23, 2018. https://on.natgeo.com/36o4QNA

Chapter 3

Augliere, Bethany. "Benchmarks: November 13, 1985: Nevado del Ruiz eruption triggers deadly lahars." Earth, November 13, 2016. https://bit.ly/3aw7ELY

Cartier, Kimberley M.S. "Honoring Volcanologist David Johnston as a Hero and a Human." Eos, June 27, 2019. https://bit.ly/2tBDrL5

Chandler, Joe. "The History of Volcanology." Sciencing, April 24, 2017. https://bit.ly/2RJuSWj

"Dec 16, 1707 CE: Last Eruption of Mount Fuji." *National Geographic*. https://bit.ly/2uwEhsj

"Hazard maps." GNS Science, Te Pū Ao. https://bit.ly/2GkgoH6

"How can we tell when a volcano will erupt?" USGS. https://on.doi.gov/3azqFx3

King, Hobart M. "Volcanic Explosivity Index (VEI)." Geology.com. https://bit.ly/38AaGNv

"Movement on the Surface Provides Information About the Subsurface." USGS. https://on.doi.gov/3azphKW

"Networks of GPS receivers track ground movement at volcanoes." USGS. https://on.doi.gov/2GiINNP

"Nevado del Ruiz." Smithsonian Institution National Museum of Natural History Global Volcanism Program. https://s.si.edu/2NUEE74

"Volcano Hazard Maps." Pacific Northwest Seismic Network. https://bit.ly/2Gi2kxZ

"Volcano Monitoring and Research." USGS. https://on.doi.gov/3aFE1ry

"Volcanoes." Weather Wiz Kids. https://bit.ly/38rqBOa

"Volcanology methods." Science Learning Hub—Pokapū Akoranga Pūtaiao. https://bit.ly/3aMM0n0

Chapter 4

Andrews, Robin. "Why Can't We Artificially Drain Magma Chambers To Stop Volcanic Eruptions?" *Forbes*, May 28, 2018. https://bit.ly/2NVhUUn

Bagley, Mary. "Mount Etna: Facts About Volcano's Eruptions." Live Science, February 28, 2017. https://bit.ly/36jlxtO

Harris, Tom. "How Volcanoes Work." HowStuffWorks. https://bit.ly/36ksScx

Hopper, David. "Greg Valentine, University at Buffalo – Volcanic Flows." The Academic Minute. https://bit.ly/2TPn1t6

JIJI. "Major Mount Fuji eruption would shower about 1.5 cm of ash on central Tokyo: disaster management team." *The Japan Times*, March 23, 2019. https://bit.ly/37mfgyR

McGuire, Bill. "How to stop a volcano." *Science Focus*, April 27, 2018. https://bit.ly/2tKXFSy

"Mt Fuji eruption scenario to be studied." The Nation Thailand, April 2, 2018. https://bit.ly/2RIBzIq

"Peach Spring Tuff exposed." The University of Arizona. https://bit.ly/2GiwHVa

Plumer, Brad. "What would happen if the Yellowstone supervolcano actually erupted." Vox, December 15, 2014. https://bit.ly/37p6eBc

Schultz, Colin. "Two Companies Want to Frack the Slopes of a Volcano." *Smithsonian Magazine*, October 4, 2012. https://bit.ly/30Pi9FR

Taylor, Alan. "The Eldfell Eruption of 1973." *The Atlantic*, January 25, 2017. https://bit.ly/37mBIrx

"Volcanic alert-levels characterize conditions at U.S. volcanoes." USGS. https://on.doi.gov/37mfkP7

"Volcano Preparedness." American Red Cross. https://rdcrss.org/37p60Km

"What do volcanoes have to do with climate change?" NASA. https://go.nasa.gov/2RG2u7A

Learning More

Books

Berg, Shannon. *Hawaii Volcano of 2018*. North Star Editions, 2019.

Law, Felicia and Gerry Bailey. *Escape from the Volcano*. Crabtree Publishing, 2016.

Maurer, Tracy Maureen Nelson. *The World's Worst Volcanic Eruptions*. Capstone Press, 2019.

Rusch, Elizabeth. *Eruption!: Volcanoes and the Science of Saving Lives*. HMH Books for Young Readers, 2017.

Websites

Learn all about how volcanoes work.
science.howstuffworks.com/nature/natural-disasters/volcano.htm

Discover how NASA scientists track volcano plumes.
earthobservatory.nasa.gov/blogs/eokids/sky-high-keeping-track-of-volcano-plumes

Find out more about the work of the USGS to study volcanoes.
https://www.usgs.gov/science-explorer-results?es=volcanoes

Watch a video of volcanoes in action.
kids.nationalgeographic.com/explore/science/volcano

Glossary

amplitude The range of something as it rises away from or dips below a calm or flat surface

collapsed Fell down or caved in

contaminated Made unclean or impure after being mixed with something else

deformation Changes in an object's shape or size due to a force being applied to the object

depression A part of the ground that is lower than the rest of the ground

depressurize To release the pressure of a gas inside an object

dormant Inactive for a long period of time

drought Long period of dryness with no rain that leads to a lack of water

economies The wealth or resources of an area or country

elements Substances that cannot be broken down into simpler substances using chemical reactions

emissions Gases released into the air

evacuate The act of moving people from a dangerous place to a safer place

extinct Not known to have erupted in recorded history

faults Cracks between Earth's crust that are found at the border between tectonic plates

fissures Narrow openings made by cracks or breaks in earth or rocks

Global Positioning System (GPS) A navigation system that can be used to find the exact location and speed of something

global warming The gradual increase in Earth's overall temperature due to human activities

humanitarian Person or group that seeks to improve the quality of life for people in need

igneous Rock that forms from lava or magma that has cooled

innovative Original or new ideas, methods, and processes

lava Hot liquid rock

magma Molten rock

mantle A layer of hot, solid rock that accounts for about 84 percent of the planet

ozone A layer of colorless gas above Earth's surface that screens harmful rays from the Sun

pyroclastic flows Hot gases and volcanic matter that typically move at high speeds from a volcano

ratio A relationship between two amounts that shows how many times one amount goes into the other

real-time Data that is sent at the same time as the event takes place

resistant Able to withstand something

satellites Objects sent into space to collect data as they orbit Earth

sea level The position of something in relation to the level of the surface of the sea

seismology The science of studying earthquakes

simultaneous Happening at the same time as something else

spectrometer Device used to measure the strength of light beams as they pass through an object

summit The highest point on a hill, mountain, or volcano

survival skills Techniques that can be used to survive in any kind of environment

tectonic Related to the movement and changing state of Earth's crust

tsunamis Massive waves that are caused by natural events

vents Openings where air, gas, and liquid can escape

Index

About the Author

Heather C. Hudak has written hundreds of books for children on topics ranging from ancient history to modern technology. When Heather is not writing, she enjoys traveling the world, and camping in the mountains with her husband and their rescue pets. She has visited many volcanoes, including Arenal Volcano in Costa Rica and Mount Bromo in Indonesia.